FARM MACHINES

SWATHER

KATHY MIDDLETON

A Crabtree Roots Book

Crabtree Publishing
crabtreebooks.com

School-to-Home Support for Caregivers and Teachers

This book helps children grow by letting them practice reading. Here are a few guiding questions to help readers build comprehension skills. Possible answers appear here in red.

Before Reading:

- What do I think this book is about?
    - *I think this book is about swathers.*
    - *I think this book will talk about what swathers do.*
- What do I want to learn about this topic?
    - *I want to learn about what jobs swathers do.*
    - *I want to learn why swathers are important farm machines.*

During Reading:

- I wonder why...
    - *I wonder why farmers use a swather.*
    - *I wonder why swathers are needed on a farm.*
- What have I learned so far?
    - *I have learned that swathers have sharp discs.*
    - *I have learned that most swathers attach to a tractor.*

After Reading:

- What details did I learn about this topic?
    - *I have learned that cut down crops on a farm.*
    - *I learned that swathers cut crops down in rows.*
- Read the book again and look for the vocabulary words.
    - *I see the words **swather** on page 3 and the word **grain** on page 10. The other vocabulary words are found on page 14.*

This is a **swather**.

Swathers cut down **crops** on a farm.

Most swathers are **vehicles**.

Swathers have sharp **discs** or a cutting bar.

Hay or **grain** is cut down to dry in the sun.

Swathers cut down crops and put them into rows.

Word List

Sight Words

a	dry	most	the
are	farm	on	this
bar	have	or	to
cut	hay	rows	
cutting	in	sharp	
down	is	sun	

Words to Know

crops

discs

grain

swather

vehicles

43 Words

This is a **swather**.

Swathers cut down **crops** on a farm.

Most swathers are **vehicles**.

Swathers have sharp **discs** or a cutting bar.

Hay or **grain** is cut down to dry in the sun.

Swathers cut down crops and put them into rows.

FARM MACHINES SWATHER

Written by: Kathy Middleton
Designed by: Under the Oaks Media
Proofreader: Crystal Sikkens
Production manager: Candice Campbell
Prepress technician: Katherine Kantor

Photographs: All images from Sutterstock

Crabtree Publishing

crabtreebooks.com 800-387-7650

 In Canada: We acknowledge the financial support of the Government of Canada through the Canada Book Fund for our publishing activities.

Printed in the U.S.A./CP2025

Published in Canada
Crabtree Publishing
616 Welland Ave.
St. Catharines, Ontario
L2M 5V6

Published in the United States
Crabtree Publishing
347 Fifth Ave
Suite 1402-145
New York, NY 10016

Library and Archives Canada Cataloguing in Publication
Available at Library and Archives Canada

Library of Congress Cataloging-in-Publication Data
Available at the Library of Congress

Hardcover: 978-1-0398-8108-2
Paperback: 978-1-0398-8469-4
Ebook (pdf): 978-1-0398-8228-7
Epub: 978-1-0398-8348-2
Audio: 978-1-0398-8511-0
Read-Along: 978-1-0398-8553-0